Jamile da Silva Oliveira
Fábio Gelape Faleiro

Passionflowers

Jamile da Silva Oliveira
Fábio Gelape Faleiro

Passionflowers

Genetic diversity and germplasm characterisation

ScienciaScripts

Imprint
Any brand names and product names mentioned in this book are subject to trademark, brand or patent protection and are trademarks or registered trademarks of their respective holders. The use of brand names, product names, common names, trade names, product descriptions etc. even without a particular marking in this work is in no way to be construed to mean that such names may be regarded as unrestricted in respect of trademark and brand protection legislation and could thus be used by anyone.

Cover image: www.ingimage.com

This book is a translation from the original published under ISBN 978-620-2-40710-6.

Publisher:
Sciencia Scripts
is a trademark of
Dodo Books Indian Ocean Ltd. and OmniScriptum S.R.L publishing group

120 High Road, East Finchley, London, N2 9ED, United Kingdom
Str. Armeneasca 28/1, office 1, Chisinau MD-2012, Republic of Moldova, Europe
Printed at: see last page
ISBN: 978-620-8-32381-3

GENETIC DIVERSITY, GERMPLASM CHARACTERIZATION AND PHYSIOLOGICAL QUALITY OF PASSIONFLOWER SEEDS

This book deals with the genetic diversity and characterization of existing germplasm in the Passifloraceae family, more specifically in the genus *Passiflora*, which is the most representative and important of the family. The genus *Passiflora* is very important in tropical fruit growing, and Brazil is the world's largest producer of maracuja, especially the sour passion fruit *(Passiflora edulis* Sims) and the sweet passion fruit *(Passiflora alata* Curtis). The country has a great diversity of genetic resources, especially species belonging to the Passifloraceae family, which has around 150 species endemic to Brazil, especially in the north-central region of the country.

The aim of this book is firstly to provide a general review of the Passifloraceae family and the genus Passiflora, and then to give the reader an overview of the state of the art of characterization work that has been and is being carried out with *Passiflora* species. Various tools used in the characterization of genetic resources involving qualitative, quantitative and molecular variables and those related to physiological quality and seed storage are exemplified, making their use simpler and more practical.

Keywords: Passifloraceae, Genetic Improvement, Resources, Genetics, Morphoagronomic characters, Descriptors for cultivar protection, Multivariate analysis, Genetic parameters, Seed physiology.

SUMMARY

1 INTRODUCTION

The genus *Passiflora* has wide intra- and interspecific genetic variability of great importance. Wild Passiflora species have interesting characteristics, such as longevity, an extended flowering period, a shorter androgynophore which facilitates pollination by smaller insects, and a higher concentration of chemical components (MELETTI et al., 2005; JUNQUEIRA et al., 2006a).

The investigation of genetic diversity is an activity of great relevance to plant breeding and the conservation of many plant species, since through this knowledge it is possible to identify contrasting materials with characteristics of interest, such as sources of resistance to diseases and high productivity, in order to make promising crosses for use in breeding programs (CRUZ and CARNEIRO, 2006).

There is concern about conserving genetic resources, but in order to know and use the conserved materials, it is essential that they are characterized and evaluated. In order for the conserved accessions to be used effectively, it is necessary to characterize them. Characterization is essential for generating knowledge about the germplasm conserved in germplasm banks and/or collections, as it allows for better management and provides support for pre-breeding and improvement programs (OLIVEIRA, 2005).

The feasibility of using genetic resources is based on the collection, introduction, conservation and exchange of germplasm accessions, as well as their characterization and evaluation, which allows us to know the qualities and potential of the material (FALEIRO et al., 2012). In a survey of research demands in passion fruit cultivation, Faleiro et al. (2006) indicated characterization as one of the priorities for passion fruit research. An important stage in this process is the evaluation and development of descriptors, which should take into account morphological, agronomic, cytological, biochemical, physiological and molecular characteristics. Regardless of the method used, the important thing is that it makes it possible to distinguish accessions, identify duplicates and accessions with characteristics of interest that can be used in breeding programs (COSTA et al., 2009). On the other hand, even when characterized, genetic variability can be threatened, since changes in allele frequencies can occur, as well as their loss or fixation, compromising the future variability of conserved accessions.

According to Vieira et al. (2007), breeders have used morphoagronomic descriptors to characterize the accessions selected during the breeding program. These descriptors play a fundamental role in the characterization and selection of plants, being decisive in the choice of genotypes throughout the recombination cycles and also in the choice of genotypes for use as new parents. Morphoagronomic characterization for genetic variability studies has been based on characters that are easy to detect and measure, and suffer little environmental influence.

Thus, throughout the germplasm conservation process, morphoagronomic descriptors and molecular markers can be used to monitor the genetic variability of accessions. The acquisition of scientific information through the characterization of *Passiflora* spp. accessions allows the valorization, conservation and use of biodiversity, with the aim of accessing new potential sources of genetic variability, guiding and increasing the efficiency of breeding programs and contributing to the development of new materials, and especially to contribute to the registration and protection processes of these new materials selected or developed in such programs.

Research into the germination and quality of Passiflora seeds is still incipient and there are many barriers to the germination process for Passiflora species, such as dormancy, which can be linked to physical factors (impermeability of the tegument to water and gases), chemical (presence of inhibitory factors), mechanical (resistance of the integument to embryo growth) or physiological (dormancy, maturity, vigor), physiological mechanisms of germination inhibition (DELANOY et al., 2006).

Seed germination occurs when, under appropriate conditions, the embryonic axis restarts its development, which had been interrupted at the time of physiological maturity (CARVALHO and NAKAGAWA, 2000). The physiological potential of seed lots is routinely assessed by the germination test, which is conducted under favorable conditions of humidity, temperature, light and substrate, allowing for the maximum expression of germination potential.

In view of the above, the aim was to review the genetic diversity and characterization of *Passiflora* germplasm based on qualitative and quantitative characteristics, molecular markers, physiological quality and seed storage.

Passionflower diversity

The genus *Passiflora* belongs to the family Passifloraceae, classified in the order Malpighiales. The family comprises 18 genera and around 630 species (MILWARD-DE-AZEVEDO and BAUMGRATZ, 2004). Souza and Lorenzi (2008) describe the Passifloraceae family as being made up of herbaceous or woody climbing plants, with tendrils originating from modifications of the inflorescences; having spiral alternate leaves, simple or rarely compound, often lobed, generally with extrafloral nectaries on the petiole or lamina, with or without stipules, entire or serrate margin.

Plants of the Passifloraceae family have a cymose or racemose inflorescence, usually reduced to a single flower, usually bisexual, actinomorphic, with a well-developed androgynophore, diclamideous or rarely monoclamideous, perigynous; calyx usually dialysespalous, imbricate pre-flower, often petaloid; corolla usually dialipetalous, preflowering imbricate; corona arranged at the apex of the hypanthium, formed by one or more cycles of appendages; stamens usually free from each other, anthers rimoseous; nectariferous disc sometimes present around the ovary or the androgynophore; ovary supero carpellar, unilocular, parietal placentation, pluriovulate, stipes usually free from each other. The fruit is a berry or capsule (SOUZA and LORENZI, 2008).

It is believed that various groups of plants were the ancestors of the genus *Passiflora*, and that they originated in regions of Africa, from where they were dispersed to Europe and Asia and diversified after arriving in Central America (MUSCHNER et al., 2012).

The genus *Passiflora* is considered to be the most representative of the Passifloraceae family, with around 500 species, most of which have their center of origin in Tropical America, of which 139 are dispersed in the Brazilian territory, placing Brazil, specifically the North Central Region of the country, among the main centers of genetic diversity of the genus (BERNACCI et al., 2008; BERNACCI et al., 2013). Ulmer and MacDougal (2004) proposed four subgenera for the genus *Passiflora,* namely *Astrophea*, with 57 species, *Deidamioides*, 13, *Decaloba*, 214 and *Passiflora*, 236. These classifications were based on morphological and ecological characters.

The genus *Passiflora* is made up of herbaceous or shrubby climbing plants, rarely erect. In general, they have cylindrical or quadrangular stems, branched, angular, suberect, glabrous or hairy (VANDERPLANK, 2000). Species of the genus *Passiflora* have enormous phenotypic variation, especially in the leaves, which can be alternate, simple or compound, entire or lobed and of variable shape, with entire or serrated margins. It is possible to observe nectariferous glands on the petiole, on the margin of the bract or on the dorsal part of the leaf (FEUILLET and MACDOUGAL, 2007; NUNES and QUEIROZ, 2007; CERVI et al., 2010).

The presence of bracts is a striking feature of most species, with the exception of some in the subgenus *Decaloba*. The position, size and shape of the bracts are important characteristics for the taxonomic separation of genera (VANDERPLANK, 2000).

Plants of the genus *Passiflora* have hermaphrodite flowers and vary greatly in shape and color, ranging from white to deep red. They have radial symmetry with a tubular, herbaceous or subcarnate calyx with five sepals. The corolla has five membranaceous petals, alternating with the sepals. The corona is usually colored and welded to the androgynophore, which are raised, which is characteristic of the Passifloraceae family (VANDERPLANK, 2000; ULMER and MACDOUGAL, 2004).

The fruits are usually indehiscent berries or dehiscent capsules with a globose or ovoid shape. They may have a yellow color, but there are also fruits with a red or purple color (VANDERPLANK, 2000; ULMER and MACDOUGAL, 2004). The skin is leathery, brittle and smooth, protecting the seeds, which are surrounded by a mucilaginous aril (BERNACCI et al., 2008; NUNES and QUEIROZ, 2007).

Genetic resources

Genetic resources involve the variability of plant, animal and microorganism species that are part of biodiversity and are of current or potential socio-economic interest for use in breeding programs, biotechnology and related areas. When the term plant genetic resources is used, the universe of biodiversity is reduced to those related to flora (NASS, 2011).

According to Faleiro et al. (2011a), plant genetic resources are a natural reservoir of genes with potential for use in the production of essential products for humanity, such as food, fibers and medicines. These resources are of great importance because they are related to basic human needs. Cultivated plants, wild species related to commercial ones, old varieties and improved varieties represent genetic resources that should be conserved and characterized, as they can be used in breeding programs.

Brazil is considered to be the center of passionflower diversity and has ample genetic variability, which is fundamental to the success of any genetic improvement program for a species (GANGA et al., 2004). The great potential of using wild passion fruit species in passion fruit breeding programs has been reported in recent years (JUNQUEIRA et al., 2006; FALEIRO et al., 2008; FALEIRO and JUNQUEIRA, 2009; FALEIRO et al., 2011a; 2011b).

The genus *Passiflora* has wide intra- and inter-species variability and has the potential for a variety of uses, both for food and medicine, as well as for ornamental purposes. Native and wild species of maracuja have potential not only for *fresh* consumption, but also for the production of pharmaceuticals, ornamental plants and functional foods. Wild species are sources of genes for the improvement of the yellow passion fruit, as rootstocks and for obtaining ornamental passion fruit hybrids (FALEIRO et al., 2006; FALEIRO et al., 2008).

Quantifying this genetic variability is fundamental for evaluating the performance of these species and thus identifying valuable genetic resources, both those that can be introduced directly into production systems and those with the potential to be used in breeding programs (FALEIRO et al., 2011a).

The conservation of different accessions in active germplasm banks (BAGs) for future genetic improvement programs with a view to better agronomic use is important, since it is a fruit tree with a wide diversity. According to Alves et al. (2005), the genetic variability existing at the population level of native species is one of the most important factors when it comes to conserving and using genetic resources in breeding programs.

The use of genetic resources is based on the collection, introduction, exchange and conservation of germplasm accessions, as well as their characterization and evaluation. Conservation provides support for genetic improvement work, enables the exchange of germplasm and, notably, the preservation of genetic variability, while characterization and evaluation make it possible to know the qualities and potential of the material (FALEIRO et al., 2011a).

The conservation of germplasm, which is an essential activity if material is to be available for breeding, can be carried out *in situ* (keeping the material in the places where it occurs) or *ex situ* (in places and conditions other than those in which it occurs naturally). *In situ* conservation aims to conserve the natural habitats in which genetic diversity exists, including native and environmentally protected areas, reserves and traditional farming systems. *Ex situ* conservation involves the removal of genetic resources from their natural environment and their transfer to artificial storage conditions, which can range from conservation in field conditions to in laboratories, for short or long periods of time (SCHERWINSKI-PEREIRA and COSTA, 2010).

Despite the importance of the genetic resources of the genus *Passiflora* for Brazil, there is a lack of research, especially in the basic areas, especially with regard to germplasm. In addition, detailed morphological, agronomic, cytogenetic and molecular characterization of all accessions is needed with a view to their practical use in commercial crops, in breeding programs, as rootstocks, in germplasm exchange and even the use of active principles, molecules and genes from this valuable genetic heritage (FALEIRO et al., 2005a; FALEIRO et al., 2011a; 2011b).

2 Characterization

Morphological characterization

For knowledge and use of conserved materials, it is essential that they are characterized and evaluated; in a survey of research demands in passion fruit cultivation, Faleiro et al. (2006) indicated the characterization and domestication of species as priority points for passion fruit research. The first stage of the characterization and evaluation process is the development of descriptors, which should take into account morphological, agronomic and molecular characteristics, among others (FALEIRO et al., 2006).

To develop a breeding program, among other things, it is first necessary to obtain and characterize the germplasm, in order to obtain basic information about the genotypes, for the practical and applied use of the products generated by the program (FALEIRO et al., 2008). The identification of intra- and interspecific variability is of essential importance, as it allows genetic resources to be used more efficiently by breeders.

Characterizing and exploring the genetic variability existing in passiflora species can reveal sources of characteristics sought in breeding programs, such as resistance or tolerance to pests and diseases (FALEIRO et al., 2006). The study of diversity is an important activity both for plant breeding and for the conservation of many species. It enables accessions to be described and differentiated, allowing contrasting genotypes to be identified in order to make promising crosses (CRUZ and CARNEIRO, 2006).

Genetic diversity expresses differences in populations, which can be estimated using various markers, such as morphoagronomic, physiological, biochemical and molecular descriptors (AMARAL JUNIOR et al., 2010).

Genetic diversity studies allow inferences to be made about the organization of germplasm, the efficiency of genotype sampling, the definition of artificial crosses and the incorporation of genes from native and exotic germplasm (VIEIRA et al., 2007).

Morphological characterization is the most accessible and widely used way of quantifying the genetic diversity of available germplasm. According to Vieira et al. (2007), breeders have used morphoagronomic descriptors to characterize accessions and plants selected during the breeding program. These descriptors play a fundamental role in the characterization and selection of plants, being decisive in the choice of a new cultivar, as well as allowing the estimation of genetic parameters and genetic diversity in accessions conserved in germplasm banks. Morphoagronomic characterization has been based on characters that are easy to detect and measure, and suffer little environmental influence.

By using morphological and agronomic or morpho-agronomic characteristics, genetic diversity can be measured. Qualitative and quantitative characters are considered easy to detect, have high heritability and suffer little environmental variation, and are used to differentiate accessions. Various multivariate analysis methods are used to quantify genetic diversity. One of these methods is the Ward-MLM *(Modified Location Model),* proposed by Franco et al. (1998). This method allows quantitative and qualitative variables to be analyzed simultaneously, accessing much of the available germplasm information. Through this technique it is also possible to define the optimum number of groups and reliably identify the best probability of each accession being allocated to a particular group (GONCALVES et al., 2009).

Tangarife et al. (2009) carried out a morphological characterization of 21 species of the genus Passiflora, including three subgenera. This study made it possible to distinguish the subgenera in a similar way to the taxonomic classification, with the variables related to the floral part contributing the most to separating the species. Viana et al. (2010) used eleven descriptors to evaluate six species of the genus Passiflora and found wide inter- and intraspecific morphological variation, clearly separating the species.

The genetic divergence of *P. cincinnata* was studied in 32 accessions conserved at the Embrapa Semi-Arido workstation in Petrolina - PE, through the observation of morphological descriptors, and was estimated based on 23 characters. The *P. cincinnata* accessions showed genetic variability for all the morphological descriptors used in the evaluation (ARAUJO et al., 2008).

Crochemore et al. (2003) found ample genetic variability when studying 55 accessions of *Passiflora* spp. using leaf, stipule, stem, tendril and fruit descriptors. Nascimento et al. (2003), aiming to select *P. edulis* progenies for fruit quality, detected great variability in the species.

Negreiros et al. (2007) also verified the existence of genetic variability between 34 half-sibling progenies and three cultivars of passion fruit (*P. edulis)*. They observed that the characteristics that most contributed to the study of genetic diversity were stem diameter at 60 days, stem diameter at the start of production and plant height at 60 days.

Chagas et al. (2016) observed the occurrence of genetic divergence between the studied populations of passion fruit, with the characteristics total fruit mass and peel mass contributing the most to the genetic diversity of passion fruit genotypes. They concluded that there is great genetic variability in the populations studied, which makes it possible to select plants with high potential for breeding purposes.

Castro et al. (2012) selected minimum morphological descriptors to characterize *P. edulis* genotypes. The result was obtained through principal component analysis, which indicated 22 of the 28 descriptors analyzed for the characterization of *P. edulis,* with a high contribution to the total variation observed.

Santos et al. (2011), using multivariate analysis, observed when quantifying the genetic diversity between *P. foetida, P. sublanceolada* and the corresponding hybrid, based on morphological characters, that the variables with the greatest contribution were flower diameter and peduncle length.

Lawinscky et al. (2014) analyzed the morphological characterization and genetic diversity of *Passiflora alata* Curtis and *P. cincinnata* Mast. and found that corona diameter was the variable that most contributed to determining the genetic diversity between the two species studied.

As morphological characterization is a fundamental stage in plant breeding programs, since it enables the germplasm to be known and allows genetic variability to be estimated, according to Paiva et al. (2014), this knowledge enables the breeder to explore diversity and introduce favourable alleles found in wild species, such as resistance to diseases, self-compatibility and reduced androgynophore, through interspecific crossing. In order for interspecific crosses to be successful, breeders must look for accessions that are genetically closer and possibly compatible. Thus, the information obtained through morphological characterization can help in the choice of genitors in passion fruit breeding programs that use interspecific hybridization.

Agronomic characterization

Agronomic characteristics can be exploited from genetic resources, especially with a view to obtaining variability that can be used to generate more productive varieties, in the search for resistance to pests and diseases, among others (FALEIRO et al., 2011a).

In agronomic characterization, many descriptors are considered, requiring more effort and time to collect the data. However, in many situations, there is no need for a large number of descriptors, and it is more rational to select those that best represent the existing variability for the crop studied. This characterization has been carried out in germplasm collections to generate information on the description and classification of the conserved material. Silva et al. (2012) estimated genetic parameters related to eleven agronomic characteristics of a population of yellow passion fruit under the recurrent selection method. The characters evaluated were number of days to flowering, average fruit weight, average fruit length, average fruit width, average peel thickness, total soluble solids content, pulp color, average pulp percentage, total number of fruits, total yield and average fruit weight. With these characteristics, the authors found that there is genetic variability available in the population, so there is a possibility of selecting superior yellow passion fruit progenies.

Sousa et al. (2012), with the aim of characterizing and studying the genetic divergence of *P. edulis* and *P. cincinnata* accessions based on agronomic descriptors, observed that the *P. edulis* and *P. cincinnata* accessions analysed showed genetic variability for most of the characteristics studied, making it possible to select divergent genitors in relation to the fruit's physical and chemical characteristics. The most important characteristics for selecting passion fruit genotypes were: number of seeds, fruit diameter, fruit size and fruit weight, with fruit size, juice yield and fruit diameter contributing the most to the total divergence between Passiflora accessions, and titratable acidity contributing the least.

Through an agronomic evaluation of yellow passion fruit parents and hybrids, Neves et al. (2013) identified at least four hybrids (H09-08, H09-10, H09-13 and H09-14) with average fruit yields above 40 t.ha^{-1} . And that some hybrids (H09-10, H09-14 and H09-20) have a good balance for the main characteristics, such as fruit yield, fruit number, fruit mass, juice yield, punch yield and total soluble solids content.

Negreiros et al. (2008) characterized the fruit of half-brother progenies of sour passion fruit in Rio Branco in the state of Acre, and observed that some of the fruit showed desirable characteristics for both the *fresh* and industrial markets, as well as the superiority of some accessions for future breeding work. They also noted that some progenies were only suitable for the *fresh* market, while others were only suitable for the industrial market. These results indicate that agronomic descriptors are useful for differentiating passiflora materials.

Understanding the genetic basis of agronomic traits is essential for breeding programs; estimates of genetic parameters, correlates and the genetic inheritance of traits are essential for determining the best breeding strategies (FALEIRO et al., 2006).

To quantify the genetic divergence between individuals, the agglomeration technique has been used, which is based on the average Euclidean distance or the generalized Mahalanobis distance (D2) (Cruz & Carneiro, 2003). Araujo et al (2008) aimed to estimate the genetic divergence between accessions of *P. cincinnata* distributed in different agro-ecological regions of the Brazilian Northeast using morpho-agronomic descriptors. They used the Tocher optimization method as a clustering method, with the generalized Mahalanobis distances (D2) as a measure of dissimilarity. They observed the existence of variability among the *P. cincinnata* accessions and that the most important characteristics for genetic divergence are: total fruit mass, pollen viability, leaf area, number of glands per bract, stem diameter, fruit and seed mass, and number of leaf glands.

Oliveira et al. (2017) aimed to characterize *Passiflora* spp. accessions and quantify genetic diversity based on quantitative morphoagronomic descriptors. The estimates were based on the Mahalanobis distance and the clustering criterion was the average linkage method between unweighted groups, UPGMA *(Unweighted Pair- Group Method using Arithmetic Avarages)*. They observed that the morpho-agronomic characterization based on quantitative descriptors of flowers and fruits contributed to the phenotypic differentiation between the 15 *Passiflora* spp. accessions, serving as an important tool for quantifying the existing variability between the accessions. The fruit characteristics made the most decisive contribution to differentiating the accessions of the species evaluated in this study. The quantitative characterization of flower and fruit structures in a complementary way is important for more complete studies of the characterization and genetic diversity of genetic resources of the genus *Passiflora*.

Molecular characterization

With the advance of modern biotechnology, tools are being created and improved to help agricultural research, such as DNA markers. These markers have applications in germplasm characterization programs, complementing the information from morphological and agronomic characteristics, increasing the power of resolution and analysis .
genetic variability of the accessions and also in the
improvement, increasing the efficiency of each stage and reducing the time spent developing new cultivars (FALEIRO et al., 2008).

The development of increasingly automated equipment and bioinformatics has led to the generation of an unlimited number of molecular markers, allowing complete coverage of the genome of interest. These advances have boosted the incorporation of molecular markers into breeding programs (PEREIRA et al., 2005). A major advantage of molecular markers is the direct investigation of genotypic characteristics, which allows the detection of variation at the DNA level, thus excluding environmental influences.

The use of molecular markers is a complementary tool for characterizing germplasm and, consequently, for identifying wild species. With regard to the study of passiflora genetic variability, the use of DNA molecular markers has been very useful as they allow a large number of genetic polymorphisms to be obtained at any stage of plant development or from cell or tissue culture (FALEIRO, 2007).

The choice of a molecular marker technique depends on its reproducibility and simplicity. The different types of molecular markers available today are differentiated by the technology used to reveal variability at the DNA level, and thus vary according to their ability to detect differences between individuals, cost, ease of use, consistency and repeatability (BOREM and CAIXETA, 2009).

Markers linked to different characteristics of economic importance have been developed (BOREM and CAIXETA, 2009). Paula et al. (2010), using molecular markers analogous to resistance genes (RGA - *Disease Resistance Gene Analogs),* evaluated wild passionflower species and found a wide variety of RGA markers. This variability makes it possible to establish degrees of genetic similarity between the materials, as well as identifying sequences related to disease resistance genes in plants, which may be useful in future crosses.

In passion fruit, molecular markers can be used efficiently in the various stages of breeding programs. In the pre-improvement stage, they can be used in the characterization and evaluation of germplasm banks, as well as in the mapping and analysis of genes of interest. In the actual breeding stage, markers can be used both in population breeding and in hybridization activities, helping to maximize not only genetic gains, but also heterosis. In the post-breeding stage, they can be used to ensure the paternity of cultivars, seminal or clonal, developed, as well as to monitor the purity of seeds or clones produced and passed on to farmers (PEREIRA et al., 2005). This is a fundamental stage in the registration and protection of cultivars.

Reis (2010) used 23 pairs of microsatellite primers to genotype 66 sour maracuja genotypes, where 25 genotypes were selected using agronomic and molecular data. The author reported that for yield, number of fruits and days to flowering, the progenies selected by molecular markers had higher averages, concluding that molecular analysis was more efficient in selecting genotypes.

However, the use of molecular markers in plant breeding is theoretically feasible in many applications. The cost of molecular marker technology, the time consumed in laboratory analysis, the cost of evaluating characteristics, among others, determines the feasibility of their use in each case (BOREM and MIRANDA, 2009).

RAPD marker

Random Amplification of Polymorphic DNA (RAPD) has proven to be efficient in identifying and quantifying genetic variability in various plant groups, which is why it has been used as an auxiliary tool in programs for characterizing and using genetic resources and breeding programs (FALEIRO, 2007). The technique is capable of detecting variations directly in DNA, and this has been used intensively for different genetic studies of various cultivars, including important work on the genetic variability of passion fruit (FALEIRO et al., 2005b).

The main advantages of the RAPD technique are the ease and speed with which the markers are obtained, the need for minimal amounts of DNA and the universalization of analyses. The main characteristics of RAPD markers are the obtaining of genomic finger*prints* of individuals, varieties and populations, structural analysis and genetic diversity in breeding populations and germplasm banks, the establishment of phylogenetic relationships between different species, the construction of genetic maps with high genomic coverage and the localization of genes of economic interest (FALEIRO, 2007).

Populations of *P. alata, P. cincinnata, P. edulis, P. foetida, P.*

gibertii, P. malacophylla, P. maliformis, P. mucronata and P. suberosa, and plants from cultivations of *P. edulis*, were evaluated with RAPD markers (VIANA et al., 2003).

Comparative studies between different passiflora species have been very successful when using RAPD molecular markers (VIANA and SOUZA, 2010). Bellon et al. (2007) evaluated divergence using RAPD markers in commercial and wild genotypes of *P. edulis* in order to direct crosses for use in the fresh, ornamental and medicinal consumer markets.

ISSR marker

The ISSR molecular marker is one of many PCR *(Polymerase Chain Reaction)* based *fingerprinting* techniques that uses simple repetitive sequence *primers* to amplify regions between target sequences. This technique has the capacity to generate a large number of multi-loci markers and can be applied to analyze practically any organism, even those for which little or no previous genetic information is available (FALEIRO, 2007). The use of ISSR markers in higher plants is known for being highly reproducible, polymorphic, informative, quick to use and inexpensive (ZIETKIEWICZ et al., 1994; BARTH et al., 2002).

ISSR *(Inter Simple Sequence Repeats)* reveals multilocus polymorphism using just one primer, anchored or unanchored, based on SSR *(Simple Sequence Repeats)* regions. The polymorphism observed can be used to infer genetic diversity and in evolutionary and taxonomic studies (REDDY et al., 2002; ISSHIKI et al., 2008).

ISSR *primers* are composed of a microsatellite sequence, usually 16 to 25 bp in length, and are used to amplify mainly Inter-SSR sequences of different lengths (FALEIRO, 2007). In contrast to other molecular markers, the target sequences of ISSR primers are abundant throughout the eukaryotic genome, which therefore helps to reveal a much larger number of polymorphic loci than other dominant markers (ANSARI et al., 2012).

The amplifiable products are generally 200-2000 bp (base pairs) long and show high reproducibility, possibly due to the use of long primers and high annealing temperatures. The limitation of this class of markers is that they are dominant, which makes it impossible to establish allelic relationships between individuals. The ISSR technique has advantages such as reproducibility, ease of use and low cost compared to many other types of molecular markers (REDDY et al., 2002).

Comparative studies between different passiflora species have been very successful using ISSR-type molecular markers (SANTOS et al., 2011). However, passiflora germplasm involving several species is still incipient. The genetic variability present in the genus *Passiflora* becomes a valuable tool for breeding efforts to find genotypes that are more suitable for intensive cultivation.

In the southwest of the state of Bahia, native populations of *P. setacea* were evaluated using 11 ISSR *primers* and four pairs of RGA (Analogs Markers of Resistance Gene) primers (PEREIRA et al., 2015). In this study, variation was detected in the percentage of polymorphic loci and considerable genetic differentiation between the populations.

Genetic parameters

Obtaining the values of genetic parameters is essential for identifying the nature of the action of the genes involved in controlling important characteristics (quantitative traits). Therefore, estimation assesses the efficiency of different breeding strategies so that there is always the possibility of gains and that genetic variability is maintained (CRUZ and CARNEIRO, 2006).

The term parameter is used to designate the characteristic constants of a population, especially the mean and the variance. Estimating genetic parameters is necessary in order to: a) obtain information on the nature of the action of the genes involved in the inheritance of the traits under investigation; and b) establish the basis for choosing the breeding methods applicable to the population. When discussing the estimation of genetic parameters, however, it must be borne in mind that the estimates obtained are only valid for the population from which the experimental material constitutes some kind of sample, and for the environmental conditions in which the study was conducted. Thus, when the aim is to experimentally estimate genetic variances, both the genotypes and the experimental environments must constitute appropriate samples of the population and the geographical area of interest, respectively (COCKERHAM, 1956).

It is also necessary to consider that it is not possible to estimate the component of genetic variation when a trial is conducted independently of the component, due to genotype x environment interaction (GARDNER, 1963). The estimation of genetic parameters is essential in quantifying the magnitude of variability and the extent to which desirable traits are inherited, in order to carry out planning with a view to promoting an efficient breeding program (VENCOVSKY and BARRIGA, 1992).

Knowing the degree of association between traits is of great value in breeding strategies, as it clarifies and quantifies the relationships between them, especially when selection for one trait promotes changes in other correlated traits (KHALIQ et al., 2004; RAMALHO et al., 2008). It is an important tool for indirect selection used when selection for a trait of interest is difficult due to low heritability or measurement and measurement problems (CRUZ and CARNEIRO, 2006). In addition, another important application of correlations is to contribute to efficiency in the simultaneous selection of traits of interest (SANTOS and VENCOVSKY, 1986).

Pre-improvement

The word *pre-improvement* has been translated from English terms such as *pre-breending, introgression breeding, genetic base broadening* or *germplasm enhancement.* It is used to define the stage in the development of germplasm into materials that are more attractive to breeders. However, the concept of pre-improvement can also vary depending on the species being studied, for example, what is pre-improvement in annual species is considered improvement in perennial species (FALEIRO et al., 2008).

According to Nass (2011), pre-improvement programs can contribute in many ways to activities related to genetic resources as well as to breeding programs. The contributions of these programs include:

- Synthesis of new base populations for breeding programs;
- Identifying potentially useful genes;
- Identification and establishment of new heterotic patterns;
- Better knowledge of the behavior of accessions *per se* and in crossbreeding;
- Increased quantity and quality of access information;
- Establishment of *core collections*;
- Greater likelihood of using plant genetic resources.

Among the possibilities mentioned above, it can be said that the generation of new base populations for improvement and the identification of heterotic patterns, which are fundamental in breeding programs that exploit hybrid vigor or heterosis, are the most tangible and significant contributions of pre-improvement programs. These programs can become important strategies for linking breeding programs with biotechnological actions, especially those dedicated to functional genomic information. In this way, pre-improvement actions could become important tools in the composition of character banks for the most varied biological fungi (NASS,

2011).

In recent years, efforts have been made to increase the genetic base of breeding programs in order to select more productive sour and sweet maracuja genotypes that are more resistant to diseases and have better physical and chemical fruit quality, One of the alternatives is pre-improvement by identifying characteristics of interest in wild species and transferring them to elite genotypes through interspecific hybridization and successive rounds of selection and recombination (FALEIRO et al., 2011b). For this reason, it is increasingly essential to carry out studies related to the characterization of wild species.

In a pre-improvement program, fundamental activities are carried out in the collections kept in germplasm banks, which are an indispensable stage, since they are aimed at identifying, characterizing and then using promising genotypes in crosses with elite germplasm (FALEIRO et al., 2011a) or even using the material directly.

The success of pre-improvement comprises at least two phases, the first referring to the knowledge of potentially useful genes or characteristics from wild species, exotic germplasm or unimproved populations; and the second, their practical use with the addition in agronomically adapted elite materials with commercial characters readily used in agricultural systems (FALEIRO et al., 2008).

Pre-improvement activities, involving the knowledge of potentially useful genes from wild species and their incorporation into varieties with commercial characteristics, are of great importance to support the practical use of genetic resources and broaden the genetic base of breeding programs. In this way, research involving the prospecting, conservation and characterization of germplasm is of strategic importance and is fundamental to enable the use of agronomic characteristics found in wild materials, such as the primary components related to productivity, fruit quality and resistance to biotic and abiotic stresses (FALEIRO et al., 2011a).

Plants of the genus *Passiflora* have agronomic and functional potential, as well as medicinal plants. Research work needs to be intensified in order to gain a better understanding of wild passion fruit germplasm (COSTA and TUPINAMBA, 2000) and can be of great help in genetic improvement programs.

Peixoto et al. (2005) describe the importance of plants of the genus *Passiflora*, which have great ornamental value due to their beautiful flowers, which are attractive due to their size, exuberant colors and original shapes, offering great prospects for exploiting their landscaping potential, as well as their fruits, which are not only eaten *fresh, but* are also used to make juices, sweets, soft drinks and ice cream.

Thus, the diversified use of passion fruit as a fruit, ornamental and medicinal plant and as a rootstock is considered an important demand for research and development actions aimed at meeting the demands of the production sector, industry and consumers (FALEIRO et al., 2006).

In order to exploit the full potential of wild passion fruit species, it is necessary to involve basic research in the areas of conservation and characterization of genetic resources and applied research aimed at genetic improvement. To maximize the success of pre-improvement, it is essential to integrate the stages with the activities and demands of breeding and post-breeding programs. Therefore, comprehensive knowledge of the intraspecific genetic variability available for improvement and market demands is essential, considering new product options and new alternatives for production systems (FALEIRO et al., 2011a).

Use of SNPC (Mapa) morphoagronomic descriptors to characterize *Passiflora* spp.

The protection of intellectual property rights relating to a cultivar is carried out through the granting of a Cultivar Protection Certificate, which is considered a movable asset for all legal purposes and the only form of protection for cultivars and rights that can prevent the free use of plants or their reproductive or vegetative multiplication parts in the country (WOLFF, 2009).

The Plant Variety Protection Act created the National Plant Variety Protection Service (SNPC) within the Ministry of Agriculture, Livestock and Food Supply (MAPA), which is responsible for managing the administrative and technical aspects of the matter. Among the various duties assigned to it are the analysis of applications and the granting of protection certificates to breeders. The SNPC is also responsible for maintaining the database and keeping live samples for inspection purposes, as well as monitoring the original characteristics of protected cultivars in the national territory (AVIANI, 2011).

In order to obtain the descriptors, the National Cultivar Protection Service (SNPC), a body linked to the Ministry of Agriculture, Livestock and Supply (MAPA), which is responsible for carrying out and monitoring cultivar protection actions in the country, has established and published a set of official instructions for carrying out distinguishability, homogeneity and stability (DHE) tests on maracuja cultivars. However, the safe and effective application of these instruments requires experimental validation with several known cultivars, in order to establish example cultivars, which are fundamental for enabling the harmonization of methodologies applied in different regions and by different evaluators (MAPA, 2017).

In a joint effort, professionals from the National Service for the Protection of Cultivars (SNPC/Mapa) and the Maracuja: germplasm and genetic improvement research group led by Embrapa, developed a set of descriptors for testing the distinctiveness, homogeneity and stability of *Passiflora edulis* Sims. This work generated a practical manual, which was designed to help improve the legal instruments for protecting cultivars of passion fruit *(P. edulis* Sims).

In order to develop this illustrated practical manual for applying *P. edulis* Sims descriptors, research was carried out to validate descriptors for characterizing sour passion fruit cultivars and to operationalize the official instructions for carrying out distinguishability, homogeneity and stability tests.

The practical manual presents 28 descriptors for branches and leaves, flowers and fruit. Each descriptor and its phenotypic classes are presented in illustrated form to standardize and avoid errors in the process of obtaining the descriptors. The morpho-agronomic descriptors used in DHE trials of sour passion fruit cultivars *(Passiflora edulis* Sims) are divided into three groups (branches-leaves, flowers and fruit), with seven descriptors for branches-leaves, 13 for flowers and eight for fruit, totaling 28 plant descriptors.

This manual is of great use to passionflower researchers, especially the teams of breeders who are working every day to develop new cultivars to meet society's demands. According to Borges et al. (2005), in order for the technological products developed by breeding programs to reach producers and benefit the entire production chain, validation and technology transfer actions are essential.

The instructions for carrying out DHE tests and the table of minimum descriptors for *P. edulis* Sims were developed and published in December 2008, in the Official Gazette of the Union No. 246 (BRASIL, 2014). In order to apply these instruments safely and effectively and to obtain the descriptors, it is important to harmonize the methodologies applied in different regions and by different evaluators.

According to Jesus et al. (2015), in order to seek the necessary harmonization for the acquisition of descriptors, an experimental validation of the descriptors used was carried out, taking into account the variations (fruit size, fruit shape, leaf shape) that occur on the same plant, the quantitative nature of some descriptors, the specific characteristics of passion fruit and the importance of using example cultivars in the evaluation of some quantitative characteristics.

Just as the practical manual for *P. edulis* Sims was developed, an illustrated practical manual for the application of descriptors was also produced for all species and interspecific hybrids of the genus *Passiflora*. This effort was made with the aim of contributing to the improvement of the legal instrument for the protection of cultivars of the different species and interspecific hybrids of the genus *Passiflora*.

According to Jesus et al. (2015), in order to obtain this practical manual, research was carried out to validate descriptors used in trials of distinctiveness, homogeneity and stability of sweet, ornamental and medicinal passion fruit cultivars, including wild species and interspecific hybrids of the *Passiflora* genus. Each descriptor and its phenotypic classes are presented in illustrated form in order to standardize and avoid errors in the process of obtaining descriptors.

As with the sour passion fruit (*P. edulis* Sims), the instructions for carrying out DHE trials and the table of minimum descriptors for sweet, ornamental and medicinal passion fruit, including wild species and interspecific hybrids *(Passiflora* spp.) were developed and published in December 2008, in the Official Gazette of the Union No. 246 (BRASIL, 2014).

The morpho-agronomic descriptors used in DHE trials of sweet, ornamental and medicinal passion fruit cultivars, including wild species and interspecific hybrids *(Passiflora* spp.), are divided into three groups (leaf-branches, flowers and fruits), with 11 descriptors for leaf-branches, 16 for flowers and eight for fruits, making a total of 35 plant descriptors.

The descriptors used in DHE trials will be useful, especially for maracuja researchers who need to collect descriptors for the registration and protection of materials developed in breeding programs. But they will also be useful for academic work and the practical differentiation of cultivars already available on the market.

Meletti et al. (2003) and Martins et al. (2003) found that within the same species there can be differences in fruit morphology such as length, diameter, pulp weight, seed, color and thickness of the peel and total soluble solids content. These intrinsic variations of the species must be taken into account in a process of adjustment and validation of the current descriptors. However, these variations are important and extremely necessary in the processes of registering and protecting cultivars, because they require a minimum degree of differentiation between the materials to be registered and protected and other materials, and even the materials that gave rise to them.

3 Physiological characterization of seeds

Numbness

Dormancy is also a problem that directly interferes with germination, and in passion fruit species it may also be related to seed storage (ALEXANDRE et al., 2004; DELANOY et al., 2006; PADUA et al., 2011). Dormancy refers to a physiological mechanism in which viable seeds do not germinate even if the right conditions are provided, such as humidity and temperature. It is an adaptive mechanism of some species that delays seed germination or leads to germination over time, increasing the probability of survival and perpetuation of the species. This dormancy mechanism is more common in wild or non-domesticated passion fruit species, since in the process of domestication, there is a selection of genotypes or processes that lead to a high percentage of germination quickly and uniformly.

As for the mechanisms of dormancy, Bewley and Black (1982) basically recognized two types: a) embryo dormancy and; b) dormancy imposed by the wrappings (or covering). The first type would include cases of metabolic inhibition and immaturity of the embryo, while the second type would include cases of impermeability of the integuments, presence of inhibitors and mechanical restriction, among others.

Dormancy can also be primary or natural (a natural characteristic of the seed) and secondary or induced (caused by unfavorable conditions for germination). Depending on the type of dormancy, there can be different causes, the most common of which are the presence of germination-inhibiting substances, the presence of physiologically immature or rudimentary embryos, the presence of an impermeable tegument, as well as a combination of the different causes (VIVIAN et al., 2008).

Depending on the cause of dormancy, there may be different

processes for overcoming dormancy, such as the use of plant regulators, chemical and mechanical scarification of the seeds, treatments with different humidity and temperature conditions, hot water, among others.

Miranda et al. (2009) reported on the current state and some advances in the sexual propagation of important species of Passifloras in Colombia, considering the histology, physiology and biochemical activities of the seeds, as well as the different methods for overcoming dormancy and storage. Considering the different species of the genus *Passiflora,* it is certainly not possible to generalize about the type of dormancy and the treatment for overcoming dormancy, and more detailed studies are needed for each species, also considering the genetic variability of different accessions and cultivars within each species.

Due to the existence of intra- and inter-species variability, it is common to find differences in germination percentage, the occurrence or not of seed dormancy, the most suitable types of treatments for overcoming dormancy, and the most suitable types of seed storage for maintaining seed viability and vigor. Some species of *Passiflora* spp. have seed dormancy due to their impermeable tegument, requiring treatment to overcome it. Ferreira (1998) observed that the seeds of some species of maracuja do not present impediments to the entry of water, although the soaking time is different for each species.

Vigor

Seed vigor is a reflection of a set of characteristics that determine its physiological potential, i.e. the ability to perform adequately when exposed to different environmental conditions. Given its importance, methods have been proposed for evaluating the physiological potential of seeds in order to estimate the safe performance of seed lots in the field and/or storage.

The impairment of vigor can vary according to genetic origin; some cultivars are more susceptible to adverse environmental conditions and have a lower capacity for development. Physiological, physiological immaturity at harvest and deterioration in storage. Morphological, the cultivar produces uneven and smaller seeds, resulting in less vigorous seedlings. Cytological, manifestation of chromosomal aberrations as a result of mutations. Mechanical, gross breakage or induction of necrosis can affect physiological mechanisms through the production of auto-toxins. Microbiotic, microorganisms present on or inside seeds during ripening can damage them under appropriate storage or field conditions by direct attack or competition for oxygen (POPINIGIS, 1977).

Seeds that are subjected to one or more of these conditions may have compromised germination and/or vigor, and therefore tend to produce weak seedlings with reduced yield potential.

Germination and seed quality of *Passiflora* spp.

Seed germination occurs when, under appropriate conditions, the embryonic axis restarts its development, which had been interrupted at the time of physiological maturation (CARVALHO and NAKAGAWA, 2000). The physiological potential of seed lots is routinely assessed by the germination test, which is conducted under optimum conditions of humidity, temperature, light and substrate, allowing for the maximum expression of germination potential.

Research in this area is still in its infancy and there are many barriers to the germination process for passiflora species, such as dormancy, which can be linked to various physical factors (impermeability of the tegument to water and gases), chemical (presence of inhibitory factors), mechanical (resistance of the integument to embryo growth) or physiological (dormancy, maturity, vigor), physiological mechanisms of germination inhibition (DELANOY et al., 2006).

Problems related to the physiological quality of seeds can be identified as the most important, such as uneven germination, which can directly compromise the formation of seedlings (NEGREIROS et al., 2006). Padua et al. (2011) also pointed out that it is important to know the other aspects that affect seed germination. These include genetic (variation between species and cultivars), pre- and post-harvest (fruit ripeness, mechanical damage during harvest, processing, phytosanitary problems, climatic variations, drying, storage), morphological and other factors.

In order to obtain high quality seeds, one of the aspects that must be taken into account is when they are collected, which can be determined by the stage of development of the fruit. With this in mind, Negreiros et al. (2006) sought to verify the influence of the stage of fruit ripeness and the post-harvest storage period on the germination and initial development of sour passion fruit *(P. edulis). edulis),* they observed that the extraction of sour passion fruit seeds should be carried out on fruit at ripeness stage 2 (fruit with 5 to 50% yellow color) and 3 (fruit with more than 50% yellow color). They also found that the development of passion fruit seedlings is best observed when the fruit is kept in storage for 3 to 6 days before the seeds are extracted.

According to Delanoy et al. (2006), several authors have shown that passion fruit germination can last from ten days to three months, with a low germination percentage and irregular seedling formation. Fowler and Bianchetti (2000) observed that some species have seed dormancy. Passion fruit seed dormancy is of the exogenous type and probably combines mechanical and chemical factors (MIRANDA et al., 2009).

Most cultivated passion fruit seeds are recalcitrant and lose their viability very quickly in the wild. Germination generally decreases when the storage period is increased. When the seeds are processed and stored, they have a low germination rate, as well as a lower germination speed and less vigor (GURUNG et al., 2014).

Research into the factors that affect viability and vigor is useful for assessing the physiological potential of seeds. It is important for defining storage strategies, especially for non-cultivated species, where the genetic and physiological heterogeneity of samples is pronounced. Given its importance, Padua et al. (2011) evaluated the germination *of P. setacea* in response to levels of dehydration, exposure to low temperatures and pre-germination treatments. These authors observed that the seeds are tolerant to desiccation up to levels close to 4% water content, that low humidity and storage temperature induce seed dormancy; and that longevity is greater when stored at subzero temperatures.

Meletti et al. (2007), evaluating the physiological quality of the seeds of six passion fruit accessions subjected to cryopreservation, observed that both at 30% and 20% humidity, the seeds of *P. nitida* behaved as recalcitrant, not accepting dehydration for storage at ultra-low temperature. Drying to 20% caused even greater damage to the seeds. After cryopreservation, in the standard germination test, most of the seeds showed burst integuments, with damage to the embryo as a result of the freezing of the structures.

The *P. edulis* and *P. ligularis* species have different levels of tolerance to desiccation, due to the sensitivity of the seeds to drying. *P. edulis* behaves as an intermediate species, unlike *P. ligularis*, which is orthodox. In addition to interspecific differences, due to differences in the sensitivity of seeds from different accessions to drying, it is suggested that there are intermediate and orthodox types within the same Passiflora species (OSPINA et al., 2000).

Germination can be promoted or accelerated by the application of growth regulators such as auxins, cytokinins, gibberellins (GAs), ethylene and others. Gibberellins are directly related to the germination of many seeds, since their exogenous application promotes the expression of genes that control the synthesis of enzymes involved in the degradation of endosperm cell walls, inducing embryo growth and stimulating the germination process. The most widely used type of gibberellin *in vitro* is gibberellic acid (GA$_3$) (BRAUN et al., 2010; SANTOS et al., 2010).

The natural germination of *P. setacea* cv. BRS Perola do Cerrado is low and very irregular, lasting from ten days to three months. In order to increase the percentage of seed germination and the uniformity of seedlings of this species, various methodologies have been tested, the most effective of which have used seed treatment with gibberellins as germination inducers (EMBRAPA, 2017a).

The natural germination *of P. cincinnata* is low and very irregular, making it necessary to use germination inducers. In the case of the genetically improved cultivar, *P. cincinnata* cv. BRS Sertao Forte, the seeds should be immersed in a solution containing 12 mL of Promalin® (GA4+7+N-phenylmethyl-aminopurine) in 988 mL of distilled water. In this solution, the concentration of the active ingredient is 225 mg per liter. Leave the seeds in the solution for 24 hours. Remove the seeds from the solution and sow them directly into plastic bags or other containers intended for sowing and seedling formation. Black polyethylene plastic bags measuring 9 cm in diameter and 15 cm in height were used (EMBRAPA, 2017b).

Seed treatment with growth regulators has significantly improved the germination percentage and uniformity of commercial cultivars of *P. setacea* cv. BRS Perola do Cerrado and *P. cincinnata* cv. BRS Sertao Forte (EMBRAPA 2017a; EMBRAPA, 2017b). In this context and due to the potential use of these species as commercial cultivars and rootstocks, more in-depth studies on seed germination and storage become extremely important for seedling production.

Comparative studies between *P. gibertii, P. cincinnata* and *P. edulis* f found that these species show great variability in terms of the percentage of germinated seeds, uniformity and speed of germination, a fact that influences seedling management in the nursery phase (VASCONCELLOS et al., 2002).

For *P. edulis* seeds, extraction of the aril by rubbing on an auger mesh sieve for three minutes, immersion in 10% quicklime for 10 minutes, rubbing on an auger mesh sieve with coarse sand for three minutes and natural fermentation for six days, promoted faster germination (MARTINS et al., 2006).

Martins et al. (2010) carried out a photochemical study of the aril of yellow maracuja seeds and observed its influence on the germination *of P. edulis* seeds. In the aril extract from the seeds of this sour maracuja, the presence of germination-inhibiting substances was observed, such as: steroids, triterpenoids, reducing sugars, which directly or indirectly interfered with water absorption.

Rosseto et al. (2000) studied the effects of pre-germination treatments *on P. alata seeds,* applying solutions of gibberellic acid. In the first study, the seeds were placed to germinate in a substrate moistened with solutions of 0, 100 and 300 ppm of gibberellic acid; in the second, the seeds were immersed in solutions of 0 and 300 ppm for 24 hours. They found that only the treatment of immersing the seeds in gibberellic acid solutions showed germination results.

Temperature also influences germination, as it is a determining factor in the speed of water absorption and, therefore, in the biochemical reactions that determine the entire process. Thus, germination will only occur within certain temperature limits, which encompass a temperature, or temperature range, in which the process occurs with maximum efficiency (CARVALHO and NAKAGAWA, 2000).

Marostega et al. (2015), studying the effect of temperature and time of immersion in water on overcoming dormancy in *P. suberosa* seeds, observed that immersing the seeds in distilled water at 50°C for 5 minutes showed the best germination percentage for the species (35%), however, the authors suggest that new analyses using other techniques to increase this germination percentage should be tested and adapted.

Passion fruit seeds show dormancy under *in vitro* conditions, causing low and uneven germination rates (JUNGHANS et al., 2006). Roncatto et al. (2006) reported that the *in* vivo germination of *P. gibertii* under a 50% shade-covered nursery provided a germination rate of 47%, which is considered low. Carvalho et al. (2012) evaluated the *in vitro* germination of *P. gibertii* with mechanical scarification and gibberellic acid. They observed that scarifying the ends of the seeds with a drip and a scalpel provided higher germination and germination speed index averages for *P. gibertii* seeds. The addition of the growth regulator GA3 to the culture medium did not affect the *in vitro* germination of *P. gibertii*.

Emergence of *Passiflora* spp. seedlings

Studies have been carried out with the aim of reducing the time needed between sowing and plant emergence and some treatments have proved to be effective in this regard. Alexandre et al. (2004), working with emergence and the emergence speed index in sour passion fruit, found that these characteristics are strongly influenced by the genotype of the plants.

Ferreira et al. (2007) evaluated the emergence of sour passion fruit seedlings from seeds treated with biostimulant and found that the seeds from the control treatment performed less well than those treated with biostimulant throughout the emergence period, with fewer seedlings emerging.

Sweet maracuja *(P. alata)* showed better seedling emergence (62; 60.5 and 57.75 %), respectively, when the aril was removed by rubbing it on a wire mesh with sand, immersing it in a solution of lime or hydrochloric acid, both with agitation for thirty minutes (OSIPI et al., 2011).

Aguiar et al. (2014) evaluated mucilage extraction methods and substrates on the emergence and development of passion fruit seedlings in a mist chamber. These authors observed that the extraction of mucilage from the seeds by washing in water or fermentation in water and the substrates carbonized rice husk and coconut fibre were the most suitable for the emergence and development of sour passion fruit plantlets.

Different external factors can negatively or positively affect the performance of seeds, including the use of biostimulants. Ferraz et al. (2014) studied the effects of biostimulants on the emergence and growth of passion fruit 'Roxinho do Kenia' *(P. edulis* Sims) seedlings. They observed that the application of biostimulant at doses of 6 and 12 $mL.kg^{-1}$ promoted an increase in the percentage of emergence.

Fresh seeds obtained from freshly harvested ripe fruit of *P. alata, P. cincinnata, P. edulis, P. gibertii* and *P. setacea* show a fast and high emergence rate after immersion in a GA3 solution of between 500 and 1,000 mg L^{-1} (SANTOS et al. 2016). As with emergence, the regulators can also provide a significant increase in plant growth, which can result in faster and more uniform seedling formation.

Plant regulators

The use of plant regulators has been advocated in fruit growing by different authors as a way of accelerating and improving seed germination and promoting the growth of young plants (Taiz and Zaiger, 2013).

Plant regulators can induce protein synthesis, promote embryo development and germination. The response of plants to plant regulators depends on many factors, including genetic and environmental ones.

Gibberellins

Gibberellins are possibly present in all plants, in all their parts and in different concentrations, with the highest concentrations in immature seeds. More than 78 gibberellins have been isolated and chemically identified. The best studied group is GA_3 (known as gibberellic acid) VIEIRA et al, 2010.

In seeds, they help overcome dormancy, promoting embryo growth and plant emergence. Specifically, gibberellins stimulate cell elongation, causing the root to break through the seed coat (TAIZ and ZEIGER, 2013).

Treatment with gibberellic acid (GA$_3$) was effective in inducing the overcoming of dormancy and accelerating the germination of *P. setacea* seeds stored at low temperatures by around 90% (PADUA et. al., 2011). According to the authors, viability, although still high, began to decline after the sixth month of storage. To prepare the GA solution$_3$, 1.5 mg of the concentrated product is diluted in 1000 ml of distilled water. After treatment, the seeds can be stored for up to 30 days at room temperature. After this period, the stored seeds begin to lose their viability.

Cytokinins

Cytokinins have many other effects on physiological development processes, including leaf senescence, nutrient mobilization, apical dominance, the formation and activity of apical meristems, floral development and seed germination (TAIZ and ZEIGER, 2013).

Cytokinins act in the opposite way to germination inhibitors, being an essential substance to complement the action of gibberellic acid in inducing germination or enzymatic processes, when these are blocked by inhibitors such as abclsic acid and/or coumarin (TAIZ and ZEIGER, 2013).

Promalin® is the commercial name given to the mixture of Gibberellin and Cytokinin, naturally occurring regulators of plant growth and ripening (Gibberellic acid No. 4 and 7 + N-(phenylmethyl)-1H- Purine-6-amine(6-benzyladeline) (Promalin - Sumimoto Chemical).

The phyto-regulator Promalin® diluted to 250 mg of the active ingredient per liter of aqueous solution, on a laboratory scale, promoted good germination of BRS Perola do Cerrado in the order of 60%. The solution is prepared with 1.66 ml of Promalin and 98.34 ml of good quality (filtered) water, and the seeds are immersed in the solution for a maximum of 20 minutes (COSTA et. al., 2010).

Storage of *Passiflora* spp. seeds

Environmental storage conditions are the most important factors for preserving seed viability, specifically temperature and relative humidity. For orthodox seeds, the best conditions for maintaining quality are low relative humidity and low temperature, as they reduce the embryo's metabolic activity and, consequently, deterioration (CARVALHO and NAKAGAWA, 2000; MARCOS FILHO, 2005).

Seed deterioration is a process that begins from the point of maximum physiological ripeness, at a progressive rate, reducing quality and culminating in the death of the seed (MARCO FILHO, 2005). Therefore, the main objective of seed storage is to reduce the rate of deterioration, since improving quality is not possible, even under ideal conditions (VILLELA and PEREZ, 2004). Successful storage depends on prior knowledge of physiological behavior during storage, since seeds of different species require special conditions for preservation.

In order to evaluate the germination of freshly harvested *P. mucronata* seeds and after storage, Santos et al. (2012) observed that freshly harvested seeds have a high germination potential, but over time germination decreases, with zero germination at four and 12 months of storage. Pre-treatment in a water bath and scarification with sandpaper promoted an increase in the germination of *P. mucronata* seeds, but not enough for those stored for 12 months.

Nogueira Filho et al. (2005), working with *P. caerulea*, observed 55% germination of freshly harvested seeds and, after 37 days of storage, seeds from the same batch as the previous one had not germinated. Meletti et al. (2002) reported that *P. setacea* seeds have a long dormancy period, requiring storage of more than two years to break dormancy. This need for storage to break dormancy was also verified for the *P. cincinnata* species (ZUCARELI et al., 2009).

Souza et al. (2105), evaluating the time and conditions of seed storage on the germination and development of *P. ligularis* (granadilla), observed that granadilla seedlings should be obtained from seeds kept in a cold room for more than 3 months. Seeds freshly removed from fruit have a low germination percentage and less growth in height when compared to plants from seeds stored for 24 months at a temperature of 7 to 10 °C.

4 FINAL CONSIDERATIONS

The maintenance of biodiversity accessions in Active Germplasm Banks is often an indispensable stage for the proper functioning of genetic improvement programs for species in general, but it is an activity that is often very costly and requires constant maintenance. The stage of characterizing the BAGs can be a very laborious stage, but it is fundamental for getting to know the conserved accessions, and through this characterization, these accessions can be effectively included in breeding programs to be used in the development of new cultivars.

Morphoagronomic characterization based on multicategorical and quantitative descriptors can contribute greatly to phenotypic differentiation between *Passiflora* spp. accessions, serving as an important tool for quantifying the variability existing in germplasm banks. It can demonstrate a clear differentiation between the species, and can be very important for more complete studies on the characterization and diversity of genetic resources in the genus *Passiflora*.

The characterization of the different structures of the plant (leaf, flower and fruit) is important for more complete studies of the characterization and genetic diversity of genetic resources of the genus *Passiflora*. It also allows characterization studies to be carried out in a fragmented way, or even in a complete way, but only accessing the information relating to the structure of interest, such as, for example, requesting only the information relating to the descriptors of the fruit.

Passiflora spp. accessions show great variation in flower and fruit characteristics, which reinforces the great diversity that exists in this genus. This variation creates more possibilities for using passifloras in breeding programs, either indirectly, by being used in crosses with the most commercially important species, *P. edulis*, or as ornamental plants, medicinal plants or for *fresh* consumption, as can be the case with species such as *P. alata, P. maliformis, P. nitida, P. edulis* (native), *P. quadrangularis, P. tenuifila* and others.

Molecular markers can contribute greatly to the characterization of accessions kept in germplasm banks. Especially if more accessible and faster techniques are used, such as ISSR and RAPD markers.

There are many types of descriptors that can be used to obtain data on the characterization of passifloras. Multicategorical qualitative descriptors, quantitative descriptors and ISSR and RAPD molecular markers are very useful tools for characterizing and studying the genetic diversity of *Passiflora* spp. accessions.

Passionflowers are very different in terms of seed storage, seed germination and seedling emergence. This diversity implies the need for different protocols for each species, especially for germination. The number of days required for germination is usually different, as is the need to use plant regulators in some species to promote or accelerate germination and emergence. Many studies are needed to determine seed germination protocols for the genus *Passiflora*.

The characterization stage of accessions contained in germplasm banks of any species is of fundamental importance because, based on characterization, genotypes can be properly evaluated and effectively included in breeding programs. Another important activity, which enables and unifies the use of information from germplasm banks, is the insertion of germplasm bank data into public consultation platforms. This type of activity should be encouraged and subsidized by the institutions that hold active germplasm banks.

The use of the descriptors developed by the SNPC (Mapa) is useful for characterizing and, above all, differentiating accessions of *Passiflora* spp. And the creation of the illustrated manuals has made it much easier to carry out the characterization required to fill in the form required by the SNPC (Mapa) for the registration and protection of cultivars.

5 BIBLIOGRAPHICAL REFERENCES

AGUIAR, R. S.; YAMAMOTO, L. Y.; PRETI, E. A.; SOUZA, G. R. B.; SBRUSSI, C. A. G.; OLIVEIRA, E. A. P.; ASSIS, A. M.; ROBERTO, S. R .; NEVES, C. S. V. J. Extraction of mucilage and substrates in the development of yellow passion fruit plantlets. **Semina: Ciencias Agrarias**, Londrina, v. 35, n. 2, p. 605-612, 2014.

ALEXANDRE, R. S.; WAGNER JUNIOR, A.; NEGREIROS, J. R. S.; PARIZZOTTO, A.; BRUCKNER, C. H. Seed germination of passion fruit genotypes. **Pesquisa Agropecuaria Brasileira**, Brasilia, v. 39, n. 12, p. 1239-1245, 2004.

ALVES, R. E.; SOUZA, F. X. de; CASTRO, A. C. R. de; RUFINO, M. do S .M.; AMARAL JUNIOR, A. T.; VIANA, A. P.; GONCALVES, L. S. A.; BARBOSA, C. D. Multivariate Procedures in Plant Genetic Resources. *In:* PEREIRA, T. N. S.(ed.). **Germplasm: Conservation, Management and Use in Plant Improvement.** Vigosa, 2010, p. 205-254.

ANSARI, S. A.; NARAYANAN, C.; WALI, S. A.; KUMAR, R.; SHUKLA, N.; RAHANGDALE, S. K. ISSR markers for analyzing molecular diversity and genetic structure of Indian teak *(Tectona grandis* Lf) populations. **Annals of Forest Research**, Romania, v. 55, n. 1, p. 1123, 2012.

ARAUJO, F. P.; SILVA, N.; QUEIROZ, M. A. Genetic divergence among *Passiflora cincinnata* Mast accessions based on morphoagronomic descriptors. **Revista Brasileira de Fruticultura**, Jaboticabal, v. 30, n. 3, p.

723-730, 2008.

AVIANI, D. M. Cultivar protection in Brazil. *In:* **Brasil. Ministry of Agriculture, Livestock and Supply. Protection of Cultivars in Brazil / Ministry of Agriculture, Livestock and Supply**. Secretariat for Agricultural Development and Cooperativism. - Brasilia: Mapa/ACS, 2011. p.27-33.

BARTH, S.; MELCHINGER, A. E.; LUBBERSTEDT, T. Genetic diversity in *Arabidopsis thaliana* L. Heynh. Investigated by cleaved amplified polymorphic sequence (CAPS) and inter-simple sequence repeat (ISSR) markers. **Molecular Ecology**, United Kingdom, v.11, n. 3, p. 495-505, 2002.

BELLON, G.; FALEIRO, F. G.; JUNQUEIRA, K. P.; JUNQUEIRA, N. T. V.; SANTOS, C. E.; BRAGA, M. F.; GUIMARAES, C. T. Genetic variability of wild and commercial maracuja *(Passiflora edulis* Sims) accessions using RAPD markers. **Revista Brasileira de Fruticultura**, Jaboticabal, v. 29, n. 1, p. 124-127, 2007.

BERNACCI, L. C.; CERVI, A. C.; MILWARD-DE-AZEVEDO, M. A.; NUNES, T. S., IMIG, D. C.; MEZZONATO, A. C. Passifloraceae. **List of Species of the Flora of Brazil**. Botanical Garden of Rio de Janeiro, 2013. Available at: <http://floradobrasil.jbrj.gov.br/>.

BERNACCI, L. C.; SOARES-SCOTT, M. D.; JUNQUEIRA, N. T. V.; PASSOS, I. R. S.; MELETTI, L. M. M. *Passiflora edulis* SIMS: The correct taxonomic way to cite the yellow pasion fruit (and of other colors). **Revista Brasileira de Fruticultura**, Jaboticabal, v. 30, n. 2, p. 566-576, 2008.

BEWLEY, J. D. and BLACK, M. 1994. **Seeds: Physiology of Development and Germination**. 2 ed. New York: Plenum Press, 1994. 445p.

BOREM, A.; MIRANDA, G. V. **Plant breeding**. 5 ed. Vigosa: UFV, 2009. 525 p.

BORGES, R. S.; SCARANARI, C.; NICOLI, A. M.; COELHO, R. R. New varieties: validation and technology transfer. *In:* FALEIRO, F. G.; JUNQUEIRA, N. T. V.; BRAGA, M. F. (Ed.) **Maracuja: germplasm and genetic improvement**. Planaltina, DF: Embrapa Cerrados, 2005. p. 619-639.

BRAZIL. Ministry of Agriculture, Livestock and Supply (MAPA). **Crop protection**. 2014. Available at :
http://www.agricultura.gov.br/vegetal/registros-autorizacoes/protecao-cultivars. Consulted on May 12, 2017.

BRAUN, H.; LOPES, J. C.; SOUZA, L. T.; SCHMILDT, E. R.; CAVATTE, R. P. Q.; CAVATTE, P. C. *In vitro* germination of beet seeds treated with gibberellic acid at different sucrose concentrations in the culture medium. **Semina: Ciencias Agrarias**, Londrina, v. 31, n. 3, p. 539-546, 2010.

CARVALHO, M. A. F.; RENATO PAIVA, R.; VARGAS, D. P.; PORTO, J. M. P.; HERRERA, R. C.; STEIN, V. C. *In vitro* germination of *Passiflora gibertii* N. E. Brown with mechanical scarification and *gibberellic acid*. Brown with mechanical scarification and gibberellic acid. **Semina:**

Ciencias Agrarias, Londrina, v. 33, n. 3, p. 10271032, 2012.

CARVALHO, N. M.; NAKAGAWA, J. **Seeds: science, technology and production**. 4. ed. Jaboticabal: FUNEP, 2000. 588 p.

CASTRO, J. A.; NEVES, C. G.; DE JESUS, O. N.; OLIVEIRA, E. J. Definition of morpho-agronomic descriptors for the characterization of yellow passion fruit. **Scientia Horticulturae**, Amsterdam, v. 145, n. 1, p. 17-22, 2012.

CERVI, A. C.; MILWARD-DE-AZEVEDO, M. A.; BERNACCI, C. Passifloraceae. **List of species of the flora of Brazil**. Rio de Janeiro Botanical Garden, 2010. Available at <http://floradobrasil.jbrj.gov.br/>.

CHAGAS, K.; ALEXANDRE, R. S.; SCHMILDT, E. R.; BRUCKNER, C. H.; FALEIRO, F. G. Genetic divergence in sour passion fruit genotypes, based on physical and chemical characteristics of the fruit. **Revista Ciencia Agronomica**, v. 47, n. 3, p. 524-531, 2016.

COCKERHAM, C. C. Effects of linkage on the covatiances between relatives. **Genetics**, Bethesda, v. 41, n. 1, p. 138-141, 1956.

COSTA, A. M.; TUPINAMBA, D. D. The maracuja and its medicinal properties - state of the art. *In:* FALEIRO, F. G.; JUNQUEIRA, N. T. V.; BRAGA, M. F. (eds.) **Maracuja: germplasm and genetic improvement**. Planaltina: Embrapa Cerrados, 2005. 475-508 p.

COSTA, C. J.; SIMOES, C. de O.; COSTA, A. M. Mechanical scarification and plant regulators for overcoming dormancy in seeds of *Passiflora setacea* D. C. Planaltina, DF: Embrapa Cerrados, 2010. 15 p. (Embrapa Cerrados. Research and development bulletin, 271).

CROCHEMORE, M. L.; MOLINARI, H. B. & STENZEL, N. M. C. Agromorphological characterization of passion fruit *(Passiflora* spp.). **Revista Brasileira de Fruticultura**, 25:5-10, 2003.

CRUZ, C. D.; CARNEIRO P. C. S. **Biometric models applied to genetic improvement**. Vigosa: UFV, 2003. v. 2. 585p.

CRUZ, C. D.; CARNEIRO, P. C. S. **Biometric models applied to genetic improvement**. v. 2, ed. 2,Vigosa: Editora UFV, 2006. 144 p.
DELANOY, M.; VAN DAMMEA, P.; SCHELDEMAN, X.; BELTRAN, J. Germination of *Passiflora mollissima* (Kunth) L H Bailey, *Passiflora tricuspis* Mast. And Passiflora nov sp. Seeds. **Scientia Horticulturae**, Amsterdam, v. 110, n. 30, p. 198-203, 2006.

EMBRAPA. **Release of the wild passion fruit cultivar BRS PeroladoCerrado**. Available at:<<http://www.cpac.embrapa.br/lancamentoperola/>. Accessed on: 27 Nov. 2017a.

EMBRAPA. **Official release of the wild passion fruit cultivar BRS Sertao Forte (BRS SF)**. Available at: <http://www.cpac.embrapa.br/lancamentosertaoforte/>. Accessed on: January 20, 2017b.

FALEIRO, F. G. **Genetic-molecular markers applied to conservation programs and the use of genetic resources**. Planaltina, DF: Embrapa

Cerrados, 2007. 102p. il.

FALEIRO, F. G.; JUNQUEIRA, N. T. V. Passion fruit *(Passiflora* spp.) improvement using wild species. *In:* MARIANTE, A. S.; SAMPAIO, M. J. A.; INGLIS, M. C. V. **The state of Brazil's plant genetic resources. Second National Report. Conservation and Sustainable Utilization for food and agriculture**. Brasilia: Embrapa Technological Information. 2009, p. 101-106.

FALEIRO, F. G.; JUNQUEIRA, N. T. V. Genetic resources: conservation, characterization and use. In: FALEIRO, F. G.; ANDRADE, S. R. M.; REIS JUNIOR, F. B. **Biotecnologia: estado da arte e aplicagoes na agropecuaria**. Planaltina: Embrapa Cerrados, 2011b. p. 513-551.

FALEIRO, F. G.; JUNQUEIRA, N. T. V.; BRAGA, M. F. Passion fruit germplasm and genetic improvement - research challenges. *In:* FALEIRO, F. G.; JUNQUEIRA, N. T. V.; BRAGA, M. F. (Eds.) **Maracuja: germplasm and genetic improvement.** Planaltina: Embrapa Cerrados, 2005a, p. 187-210.

FALEIRO, F. G.; JUNQUEIRA, N. T. V.; BRAGA, M. F. **Maracuja: demands for research.** Planaltina, DF: Embrapa Cerrados, 2006, 54p. FALEIRO, F. G.; JUNQUEIRA, N. T. V.; BRAGA, M. F. Maracuja: potential of wild passion fruit species as a source of resistance to diseases. *In:* FALEIRO, F.G., JUNQUEIRA, N.T.V.; BRAGA, M.F. (Eds.). **Passion fruit germplasm and genetic improvement - research challenges**. Planaltina, DF: Embrapa Cerrados. p.81-107. 2005b.

FALEIRO, F. G.; JUNQUEIRA, N. T. V.; BRAGA, M. F.; COSTA, A. M.

Conservation and characterization of wild passion fruit species *(Passiflora* spp.) and potential use in genetic improvement, as rootstocks, functional foods, ornamental and medicinal plants - research results. Planaltina: Embrapa Cerrados, 2012 (Documentos, Nº 312). 34p.

FALEIRO, F. G.; JUNQUEIRA, N. T. V.; BRAGA, M. F.; PEIXOTO, J. R. Pre-improvement of maracuja. *In:* LOPES, M. A.; FAVERO, A. P.; FERREIRA, M. A. J. F.; FALEIRO, F. G.; FOLLE, S. M.; GUIMARAES, E. P. (Eds.) **Plant breeding: state of the art and successful experiences**. Brasilia: Embrapa Informagao Tecnologica, 2011a. 550-570p.

FALEIRO, F. G.; JUNQUEIRA, N. T. V.; BRAGA, M. F.; PEIXOTO, J. R. **Germplasm characterization and genetic improvement of passion fruit assisted by molecular markers: research results 2005-2008**. Planaltina: Embrapa Cerrados, (Research and Development Bulletin, No. 207), 2008. 59 p.

FALEIRO, F. **Molecular markers applied to conservation programs and the use of genetic resources**. Planaltina: Embrapa Cerrados, 2007. 102 p.

FERRAZ, R. A.; SOUZA, J. M. A.; SANTOS, A. M. F.; GONCALVES, B. H. L.; REIS, L. L.; LEONEL, S. Effects of biostimulants on the emergence of 'Roxinho do Kenia' passion fruit plants. **Bioscience Journal**, Uberlandia, v. 30, n. 6, p. 1787-1792, 2014.

FERREIRA, E. G. **Produgao de frutiras nativas**. Fortaleza: Instituto Frutal, 2005. 213 p.

FERREIRA, G. Study of soaking and the effect of phytoregulators on the germination of Passifloraceae seeds. **Thesis** (Doctorate in Horticulture), Faculty of Agronomic Sciences, São Paulo State University, Botucatu, 1998.

FERREIRA, G.; COSTA, P. N.; FERRARI, T. B.; RODRIGUES, J. D.; BRAGA, J. F.; JESUS, F. A. Emergence and development of passion fruit plants from seeds treated with biostimulants. **Revista Brasileira de Fruticultura**, Jaboticabal, v. 29, n. 3, p. 595-599, 2007.

FEUILLET, C.; MACDOUGAL, J. M. Passifloraceae. *In:* KUBITZI, K. **The Families and Genera of Vascular Plants**. Berlin: Springer, v. IX, p. 270-281, 2007.

FOWLER, J. A. P.; BIANCHETTI, A. **Dormancy in forest seeds**. Colombo: Embrapa Florestas, 2000. 27 p.

FRANCO, J.; CROSSA J.; VILLASENOR, J.; TABA, S. Classifying genetic resources by categorical and continuous variables. **Crop Science**, Madison, v. 38, n.6, p.1688-1696, 1998.

GANGA, R. M. D.; RUGGIERO, C.; LEMOS, E. G. M.; GRILI, G. V. G.; GONCALVES, M. M.; CHAGAS, E. A.; WICKERT, E. Genetic diversity in yellow passion fruit using AFLP molecular markers. **Revista Brasileira de Fruticultura**, Jaboticabal, v. 26, n. 3, p. 494498, 2004.

GARDNER, O. G. Estimates of genetic parameters in cross fertilizing plants and their implications in plant breeding. In: Hanson, W.D.; Robinson, H.F. (Eds). **Statistical genetics and plant breeding**. Washington: National

Academy of Science, 1963. p. 225-252.

GONCALVES, L. S.; RODRIGUES, R.; DO AMARAL JUNIOR, A T.; KARASAWA, M.; SUDRE, C. P. Heirloom tomato gene bank: assessing genetic divergence based on morphological, agronomic and molecular data using a Ward-modified location model. **Genetics and Molecular Research**, Sao Paulo, v. 8, n.1, p. 364-374, 2009.

GURUNG, N.; SWAMY, G. S. K.; SARKAR, S. K.; BHUTIAAND, S. O.; BHUTIA, K. C. Studies on seed viability of passion fruit *(Passiflora edulis* f. flavicarpa Deg.). **Journal of Crop and Weed**, Nadia, v. 10, n. 2, p. 484487, 2014.

ISSHIKI, S.; IWATA N.; KHAN, M.M.R. ISSR variations in eggplant *(Solanum melongenal)* and related Solanum species. **Science horticulturae**, Amsterdam, v. 117, n. 3, p. 186-190, 2008.

JESUS, O. N.; MARTINS, C. A. D.; MACHADO, C. F.; OLIVEIRA, E. J.; SOARES, T. L.; FALEIRO, F. G. et al. **Application of morphoagronomic descriptors used in DHE trials of sweet, ornamental and medicinal passion fruit cultivars, including wild species and interspecific hybrids *(Passiflora* spp.).** Practical manual. Ed. I, Brasilia- DF: Embrapa, 45 p. 2015.

JUNGHANS, T. G.; VIANA, A. J. C.; JUNGHANS, D. T. In vitro and ex vitro germination of maracuja gibertii seeds with and without partially removed tegument. In: Congresso brasileiro de fruticultura, 19, 2006, Cabo Frio. **Proceedings**... Cabo Frio: SBF/UENF/ UFRuralRJ, 2006. 191 p.

JUNQUEIRA, N. T. V.; BRAGA, M. F.; FALEIRO, F. G.; PEIXOTO, J. R.; BERNACCI, L. C. Potential of wild passion fruit species as a source of resistance to diseases. In: FALEIRO, F. G.; JUNQUEIRA, N. T. V.; BRAGA, M.F. (Eds.) **Maracuja: germplasm and genetic improvement**. Planaltina: Embrapa Cerrados, 2005, p.81-108.

JUNQUEIRA, N. T. V.; FALEIRO, F. G.; BRAGA, M. F.; PEIXOTO, J. R. Use of wild *Passiflora* species in the pre-improvement of passion fruit. *In:* LOPES, M. A.; FAVERO, A. P.; FERREIRA, M. A. J. F.; FALEIRO, F. G. (Eds.) **International course on plant breeding**. Brasilia: Embrapa, 2006, p.133-137.

JUNQUEIRA, N. T. V.; LAGE, D. A. C.; BRAGA, M. D.; PEIXOTO, J. R.; BORGES, T. A.; ANDRADE, S. R. M. Reaction to diseases and productivity of a clone of sour passion fruit propagated by cuttings and grafting on herbaceous cuttings of wild passiflora. **Revista Brasileira de Fruticultura**, Jaboticabal, v. 28, n. 1, p. 97-100, 2006 a.

KHALIQ, I.; PARVEEN, N.; CHOWDHRY, M. A. Correlation and path coefficient analyses in bread wheat. **International Journal of Agriculture & Biology**, Pakistan, v. 6, n. 4, p. 633-635, 2004.

LAWINSCKY, P. R.; et al. Morphological characterization and genetic diversity in *Passiflora alata* Curtis and *P. cincinnata* Mast. (Passifloraceae) **Brazilian Journal Botany**, 37(3):261-272, 2014.

MAPA. **Crop protection**. Available at: <http://www.agricultura.gov.br/vegetal/registros-autorizacoes/protecao-cultivares>. Accessed on: June 19, 2017.

MARCOS FILHO, J. **Fisiologia de sementes de plantas cultivadas**. Piracicaba: FEALQ, 2005. 495 p.

MAROSTEGA, T. N.; CUIABANO, M. N.; RANZANI, R. E.; LUZ, P. B.; PAIVA SOBRINHO, S. Effect of heat treatment on overcoming dormancy in seeds of *Passiflora suberosa* L. **Bioscience Journal**, Uberlandia, v. 31, n. 2, p. 445-450, 2015.

MARTINS, C. M.; VASCONCELLOS, M. A. S.; ROSSETTO, C. A. V.; CARVALHO, M. G. Photochemical investigation of the aril of yellow passion fruit seeds and its influence on seed germination. **Ciencia Rural**, Santa Maria, v. 40, n. 9, p.1934-1940, 2010.

MARTINS, M. R.; REIS, M. C.; MENDES NETO, J. A.; GUSMAO, L. L.; GOMES, J. J. A. Influence of different methods of removing the aril on the germination of yellow passion fruit seeds *(Passiflora edulis* Sims f. flavicarpa Deg.). **Revista da Faculdade de Zootecnia, Veterinaria e Agronomia**, Uruguaiana, v. 13, n. 2, p. 28-38, 2006.

MARTINS, M.R.; OLIVEIRA, J.C.; DI MAURO, A.R.; SILVA, P.C. Evaluations of populations of sweet passion fruit (Passiflora alata Curtis) obtained from open populations. **Revista Brasileira de Fruticultura**, Jaboticabal, v.25, p.111-114, 2003.

MELETTI, L. M. M.; BARBOSA, W.; VEIGA, R. F. A.; PIO, R. Cryoconservation of seeds from six passion fruit accessions. **Scientia**

Agraria Paranaensis, Marechal Candido Rondon, v. 6, n. 1-2, p. 13- 20, 2007.

MELETTI, L. New technologies improve the production of maracuja seedlings. **O Agronomico**, Campinas, n. 54, p. 30-33, 2002.

MELETTI, L.M.M.; BERNACCI, L.C.; SOARES-SCOTT, M.D.; AZEVEDO FILHO, J.A.; MARTINS, A.L.M. Genetic variability in morphological, agronomic and cytogenetic characters of sweet passion fruit (Passiflora alata Curtis) populations. **Revista Brasileira de Fruticultura**, Jaboticabal, v.25, p. 275-278, 2003.

MILWARD-DE-AZEVEDO, M. A., BAUMGRATZ, J. F. A. *Passiflora* L. subgenero *decaloba* (DC.) Rchb. (Passifloraceae) in the Southeast Region of Brazil. **Rodriguesia**, Rio de Janeiro, v. 55, n. 85, p. 17-54, 2004.

MIRANDA, D.; PEREA, M.; MAGNITSKIY, S. Propagation of *Pasifloraceous* Species. *In:* MIRANDA, D.; FISCHER, G.; CARRANZA, C.; MAGNITSKIY, S.; CASIERRA, F.; PIEDRAHITA, W.; FLOREZ, L. E. **Cultivo, poscosecha y comercializacion de las pasifloraceas en Colombia: maracuya, granadilla, gulupa y curuba**. Eds. Bogota: Sociedad Colombiana de Ciencias Horticolas. 2009, p. 69-96.

MUSCHNER, V. C., ZAMBERLAN, P. M., BONATTO, S. L., FREITAS, L. B. Phylogeny, biogeography and divergence times in *Passiflora* (Passifloraceae). **Genetics and Molecular Biology**, Ribeirao Preto, v.35, n. 4, p. 1036-1043, 2012.

Nascimento WMO do, Tome AT, Oliveira M do SP de & Carvalho JEU de (2003) Selection of yellow passion fruit *(Passiflora edulis* f. *flavicarpa)* progenies for fruit quality. **Revista Brasileira de Fruticultura**, 25:186-188.

NASS, L. L. Plant breeding. In: LOPES, M. A.; FAVERO, A. P.; FERREIRA, M. A. J. F.; FALEIRO, F. G.; FOLLE, S. M.; GUIMARAES, E. P. (Eds.) **Plant breeding: state of the art and successful experiences.** Brasilia: Embrapa Informagao Tecnologica, 2011. p. 25-38.

NEGREIROS, J. R. S.; ARAUJO NETO, S. E.; ALVARES, V. S.; LIMA, V. A.; OLIVEIRA, T. K. Fruit characterization of half-firm progenies of yellow passion fruit in Rio Branco - Acre. **Revista Brasileira de Fruticultura**, Jaboticabal, v. 30, n. 2, p. 431-437, 2008.

NEGREIROS, J. R. S.; BRUCKNER, C. H.; CRUZ, C. D.; ALVARES, V. S.; MORGADO, M. A. D.; SIQUEIRA, D. L. Genetic diversity among yellow passion fruit progenies based on morphological and agronomic characteristics. **Revista Ceres**, v. 54, n. 312, p. 153-160, 2007.

NEGREIROS, J. R. S.; WAGNER JUNIOR, A.; ALVARES, V. S.; SILVA, J. O. C.; NUNES, E. S.; ALEXANDRE, R. S.; PIMENTEL, L. D.; BRUCKNER, C. H. Influence of ripening stage and post-harvest storage on germination and initial development of yellow passion fruit. **Revista Brasileira de Fruticultura**, Jaboticabal, v. 28, n. 1, p. 21-24, 2006.

NEVES, C. G.; JESUS, O. N.; LEDO, C. A. S.; OLIVEIRA, E. J. Agronomic evaluation of yellow passion fruit parents and hybrids. **Revista Brasileira Fruticultura**, Jaboticabal, v. 35, n. 1, p. 191-198, 2013.

NOGUEIRA-FILHO, G. C.; RONCATTO, G.; RUGGIERO, C.; OLIVEIRA, J. C.; MALHEIROS, E. B. Vegetative propagation of passion fruit - Achieving new accessions. *In:* FALEIRO, F. G.;

JUNQUEIRA, N. T. V.; BRAGA, M. F. **Maracuja: germplasm and genetic improvement**. Planaltina: Embrapa Cerrados, 2005. p. 341-383.

NUNES, T. S., QUEIROZ, L. P. A new species of Passiflora L. (Passifloraceae) for Brazil. **Acta Botanica Brasilica**, Belo Horizonte, v. 21, n. 2, p. 499-502, 2007.

OLIVEIRA, J. S.; FALEIRO, F. G.; JUNQUEIRA, N. T. V.; VIANA, M. L. Genetic and morphoagronomic diversity of *Passiflora* spp. based on quantitative measurements of flowers and fruits. Revista Brasileira de Fruticultura, v. 39, n. 1: (e-003), 2017.

OSIPI, E. A. F.; LIMA, C. B.; COSSA, C. A. Influence of aril removal methods on the physiological quality of *Passiflora alata* Curtis seeds. **Revista Brasileira de Fruticultura**, Jaboticabal, (Special volume), p. 680-685, 2011.

OSPINA, J. A.; GUEVARA, C. L.; CAICEDO, L. E.; BARNEY, V. Effects of moisture on Passiflora seed viability after immersion in liquid nitrogen. In: ENGELMANN, F.; HIROKO, T. (eds.). **Cryopreservation of Tropical Plant Germplasm: Current research progress and application**

Japan International Research Center for Agricultural Sciences. Tsukuba. 2000. p. 384-388.

PADUA, J. G.; SCHWINGEL, L. C.; MUNDIM, R. C.; SALOMAO, A. N.; ROVERIJOSE, S. C. B. Germination of *Passiflora setacea* seeds and dormancy induced by storage. **Revista Brasileira de Sementes**, Brasilia, v. 33, n. 1 p. 080-085, 2011.

PAIVA, C. L.; VIANA, A. P.; SANTOS, E. A.; SILVA, R. N. O.; OLIVEIRA, E. J. Genetic diversity of *Passiflora* species using the Ward-MLMl strategy. **Revista Brasileira Fruticultura**, Jaboticabal, v. 36, n. 2, p. 381 - 390, 2014.

PAULA, M. S.; FONSECA, M. E. N.; BOITEUX, L. S.; PEIXOTO, J. R. Genetic characterization of Passiflora species by molecular markers analogous to resistance genes. **Revista Brasileira de Fruticultura**, Jaboticabal, v. 32, n. 1, p. 222-229, 2010.

PEIXOTO, M. Problemas e perspectivas do maracuja ornamental. In: FALEIRO, F. G.; JUNQUEIRA, N. T. V.; BRAGA, M. F. **Maracuja: germplasm and genetic improvement**. Planaltina: Embrapa Cerrados, 2005, p. 457-463.

PEREIRA, D. A.; CORREA, R. X.; OLIVEIRA, A. C. Molecular genetic diversity and differentiation of populations of 'somnus' passion fruit trees *(Passiflora setacea* DC): Implications for conservation and pre-breeding. **Biochemical Systematics and Ecology**, Amsterdam, v. 59, n. p. 12-21, 2015.

PEREIRA, K. J. C.; DIAS, D. C. F. Germination and vigor of yellow maracuja seeds *(Passiflora edulis* Sims f. *flavicarpa* Deg.) submitted to different mucilage removal methods. **Revista Brasileira de Sementes**, Brasilia, v.22, n. 1, p.288-291, 2000.

PEREIRA, M. G.; PEREIRA, T. N. S.; PIO VIANA, A. Molecular markers applied to passion fruit improvement. In: FALEIRO, F. G.; JUNQUEIRA, N. T. V.; BRAGA, M. F. (Eds). **Maracuja- germplasm and genetic improvement**. Planaltina: Embrapa Cerrados, 2005, p. 277-292.

POPINIGIS, F. **Seed Physiology**. Brasilia: Agiplan, 1977, 297p.

RAMALHO, M. A. P.; SANTOS, J. B.; PINTO, C. B. **Genetica na agropecuaria**. 4. ed. Lavras: UFLA, 2008. 463 p.

REDDY, P. M.; SARLA, N.; ANDSIDDIQ, E. A. Inter simple sequence repeat (ISSR) polymorphism and its application in plant breeding. **Euphytica**, Netherlands, v. 128, n. 1, p.9-17, 2002.

REIS, R. V.; OLIVEIRA, E. J.; VIANA, A. P.; PEREIRA, T. N. S. Genetic diversity in recurrent selection of yellow passion fruit detected by microsatellite markers. **Pesquisa Agropecuaria Brasileira**, Brasilia, v. 46, n. 1, p. 51-57, 2011.

RONCATTO, G.; LENZA, J. B.; VALENTE, J. P.; FERREIRA, L. G.; DAMASCENO, M. A. P. Evaluation of the germination of native passion fruit species. In: Congresso Brasileiro de Fruticultura, 19, Cabo Frio. **Proceedings**... Cabo Frio: SBF/UENF/ UFRural-RJ, 2006, p.165.

ROSSETO, C. A. V.; CONEGLIAN, R. C. C.; NAKAGAWA, J. Germination of sweet maracuja *(Passiflora alata* Dryand) seeds as a function of pre-germination treatment. **Revista Brasileira de Sementes**, Brasilia, v. 22, n. 1, p. 247-252, 2000.

SANTOS, C. H. B.; CRUZ NETO, A. J.; JUNGHANS, T. G.; JESUS, O. N.; GIRARDI, E. A. Fruit ripening stage and influence of gibberellic acid on the emergence and growth of *Passiflora* spp. **Revista Ciencia Agronomica**, Fortaleza, v. 47, n. 3, p. 481-490, 2016.

SANTOS, E. A.; SOUZA, M. M.; VIANA, A P.; ALMEIDA, A. A. F.; FREITAS, J. C. O.; LAWINSCKY, P. R. Multivariate analysis of morphological characteristics of two species of passion flower with ornamental potential and of hybrids between them. **Genetics and Molecular Research**, Sao Paulo, v.10, p. 2457-2.471, 2011.

SANTOS, F. C.; RAMOS, J. D.; PASQUAL, M.; REZENDE, J. C.; SANTOS, F. C.; VILLA, F. Micropropagation of passion fruit. **Revista Ceres**, Vigosa, v. 57, n. 1, p. 112-117, 2010.

SANTOS, J.; VENCOVSKY, R. Phenotypic and genetic correlation between some agronomic characters of the bean *plant (Phaseolus vulgaris L.)*. **Ciencia e Pratica**, Lavras, v.10, n.3, p.265-272, 1986.

SANTOS, L. F.; OLIVEIRA, E. J.; SANTOS, A. S.; CARVALHO, F. M.; COSTA, J. L.; PADUA, J. G. ISSR markers as a tool for the assessment of genetic diversity in Passiflora. **Biochemical Genetics**, San Francisco, v. 49, n. 7-8, p. 540-554, 2011.

SANTOS, T. M.; FLORES, P. S.; OLIVEIRA, S. P.; SILVA, D. F. P.; BRUCKNER, C. H. Storage time and dormancy breaking methods in maracuja-de-restinga seeds. **Revista Brasileira de Agropecuaria Sustentavel**, Vigosa, v. 2, n. 1, p. 26-31, 2012.

SCHERWINSKI-PEREIRA, J. E.; COSTA, F. H. S. *In vitro* conservation of plant genetic resources: strategies, principles and applications. In: BARRUETO CID, L. P. (Org.). *In vitro* **plant cultivation**. Brasilia: Embrapa Informagao Tecnologica, 2010, p. 177-234.

SILVA, M. G. M.; VIANA, A. P.; AMARAL JUNIOR, A. T.; GONCALVEZ, L. S. A.; REIS, R. V. Biometry applied to intrapopulation improvement of yellow passion fruit. **Revista Ciencia Agronomica**, Fortaleza, v. 4 3, n. 3, p. 493-499, 2012.

SIQUEIRA, D. L.; PEREIRA, W. E. Propagation. In: BRUCKNER, C. H.; PICANCO, M. C. **Maracuja: production technology, post-harvest, agro-industry, market**. Porto Alegre: Editora Cinco Continentes, 2001, p. 85-137.

SOUSA, L. B.; SILVA, E. M.; GOMES, R. L. F.; LOPES, A. C. A.; SILVA, I. C. V. Characterization and genetic divergence of *Passiflora edulis* and *P. cincinnata* accessions based on fruit physical and chemical characteristics, **Revista Brasileira de Fruticultura**, Jaboticabal, v. 34, n. 3, p. 832-839, 2012.

SOUZA, A. D.; AOYAMA, E. M.; FURLAN, M. R. Time and storage conditions of seeds on germination and seedling development

Passiflora ligularis Juss. **Journal of Agribusiness and the Environment**, Maringa, v. 8, n. 1, p. 181-192, 2015.

SOUZA, V. C.; LORENZI, H. **Botanica Sistematica: guia ilustrado para identificagao das famflias de fanerogamas nativas e exoticas no Brasil, baseado em APG II.** 2° Ed. Nova Odessa: Instituto Plantarum, 2008, 640 p.

TAIZ, L. & ZEIGER, E. (2013). **Plant physiology** (5ed., 954p). Porto Alegre: Artmed.

TANGARIFE, M. M. M.; CAETANO, C. M.; TIQUE, C. A. P. Caracterizacion morfologica de especies del genero Passiflora de Colombia. **Acta Agronomica**, Palmira, v.58, n.3, p.117-125, 2009.

ULMER, T.; MACDOUGAL, J. M. **Passiflora: Passionflowers of the World.** Portlad-Cambridge: Timber Press. 2004, 430 p.

VANDERPLANK, R. J. R. **Passion flowers.** 3ª ed. Cambridge: The MIT Press. 2000, 224 p.

VASCONCELLOS, M. A. S.; SILVA, A. C., SILVA, A. C.; REIS, F. O. Passion fruit ecophysiology and implications for diversified exploitation. *In:* FABIO G. F.; JUNQUEIRA, N. T. V.; BRAGA, M. F. (Eds.), **Maracuja: Germplasm and genetic improvement**. Planaltina:
Embrapa Cerrados. 2005, p. 295-313.

VENCOVSKY, R.; BARRIGA, P. **Genetica biometrica no fitomelhoramento**. Ribeirao Preto: Sociedade Brasileira de Genetica, 1992, 496 p.

VIANA, A. J. C.; SOUZA, M. M.; ARAUJO, I. S.; CORREA, R. X. Genetic diversity in Passiflora species determined by morphological and molecular characteristics. **Biologia Plantarum**, Praha, v.54, n.3, p.535538, 2010.

VIANA, A. P.; PEREIRA, T. N. S.; PEREIRA, M. G.; SOUZA, M. M.; MALDONADO, F.; AMARAL JUNIOR, A. T. Diversity among genotypes of yellow passion fruit *(Passiflora edulis* f. flavicarpa) and among species of passifloras determined by RAPD markers. **Revista Brasileira de**

Fruticultura, Jaboticabal, v. 25, n. 3, p. 489-493, 2003.

VIEIRA, E. A.; CARVALHO, F. I. F.; OLIVEIRA, A. C.; MARTINS, L. F.; BENIN, G.; SILVA, J. A. G.; KOPP, M. M.; HARTWIG, I.; CARVALHO, M. F.; VALERIO, I. P. Association of genetic distance in wheat estimated from morphological characters, phenological characters and grain yield components. **Revista Brasileira de Agrociencia**, Pelotas, v. 13, n. 2, p. 161-168, 2007.

VILLELA, F. A.; PEREZ, W. B. Seed technology - collection, processing and storage. In: FERREIRA, A. G. E BORGHETTI, F. (Coord.) **Germination - from basic to applied.** Porto Alegre: Artmed, 2004. p. 265-280.

VIVIAN, R.; SILVA, A.A.; GIMENES, Jr., M.; FAGAN, E.B.; RUIZ, S.T.; LABONIA, V Dormancy in weed seeds as a survival mechanism - brief review. **Planta Daninha**, v.26, n.3, p.695-706, 2008.

WOLFF, S. **Subsidios ao IV Relatorio Nacional para a Convengao sobre Diversidade Biológica - CDB: Diagnostico sobre a Legislagao Ambiental Brasileira**. MINISTRY OF THE ENVIRONMENT. Secretariat for Biodiversity and Forests. Department of Biodiversity Conservation. 2009, 120p.

ZAIDAN, L. B. P.; BARBEDO, C. J. Breaking seed dormancy. In: FERREIRA, A. G.; BORGHETTI, F. **Germination: from basic to applied.** Porto Alegre: Artmed. 2004, p. 135-146.

I want morebooks!

Buy your books fast and straightforward online - at one of world's fastest growing online book stores! Environmentally sound due to Print-on-Demand technologies.

Buy your books online at
www.morebooks.shop

Kaufen Sie Ihre Bücher schnell und unkompliziert online – auf einer der am schnellsten wachsenden Buchhandelsplattformen weltweit! Dank Print-On-Demand umwelt- und ressourcenschonend produzi ert.

Bücher schneller online kaufen
www.morebooks.shop

info@omniscriptum.com
www.omniscriptum.com

Printed by Books on Demand GmbH, Norderstedt / Germany